Surviving a Math Class

From a **Straight A** Grad Student

Eve Wallis

Copyright © 2021 Eve Wallis Designs

HTTPS://EVEWALLIS.WORDPRESS.COM

This material, content, and format are intellectual property of Eve Wallis of "Eve Wallis Designs^{TM}". All rights reserved.

No part of this material can be reproduced by any means, or transmitted, or translated into electronic format without the written permission of Eve Wallis.

2021, First Edition

Cover, chapter, and book art created with Canva:

- Science Formula and math equation abstract background by monsitj
- Lighthouse Tower Building and Water Ocean by Weasley99
- Sailboat Conveyance Transport Isolated Icon by iconsy
- Alien Peace Sign Gesture by Clker-Free-Vector-Images
- A Complicated Circular Maze Labyrinth by Oleksa
- Additional image credits provided within illustrations

Copy editing by Eve Wallis.

ISBN: 978-1-7372820-2-0

Contents

Preface   1

1 Planning Ahead   3
    Who is the instructor?   4
    What is the teaching style?   5
    What resources will be required?   6
    What if there is no textbook?   7
    Where will you get support?   8

2 Opening Your Mind   11
    Take an active approach.   12
    Be open to change.   13
    Avoid the anxiety bug.   15

3 Studying Smart   19
    Where are you studying?   21
    What do you need when you study?   22
    How are you studying?   23
    How can you stay motivated?   24

4 Using a Textbook   27
    Using the definitions.   28
    Using the examples.   30
    Using the exercises.   32
    Being an active reader.   33

Contents

5 Learning the Big Ideas   **37**
    Minimizing necessary memorization 38
    Maximizing retention. 40
    Making mind maps. 41

6 Showing Your Work   **45**
    Show work to save time. 46
    Show work to get help. 47
    Show neat work. 47
    Show work for test success. 49

7 Solving Problems   **51**
    Solving Word Problems 52
    Finding Careless Math Errors 54
    Help! I am stuck! 55

8 Asking for Help   **59**
    Before getting help 61
    While getting help 62
    After getting help 62

9 Testing Strategies   **65**
    Before taking a test 66
    While taking a test 70
    After taking a test 71
    When you have test anxiety 73

10 Learning Remotely   **77**
    How you should participate. 78
    How to complete assignments. 80
    How you should study. 81
    How to take online exams. 82

11 Avoiding Burnout 87
Taking math with lighter classes. 88
Not procrastinating in math. 89
Not making math too hard 89

12 Worst Case Scenarios 93
Participation Scenarios 94
Assignment Scenarios 98
Exam Scenarios 102

Appendix 108
Worst Case Scenario Solutions 109
16 Things Active Learners Do 112
15 Things Active Readers Do 113
Test Preparation Tips 114
Test Taking Tips 115
Common Math Errors 116
Common Test Taking Errors 117
Help! I am stuck! Tips 118

Preface

"There is a no royal road
to anything,
one thing at a time,
all things in succession.
That which grows fast,
withers as rapidly.
That which grows slowly,
endures."

-Josiah Gilbert Holland

There is no shortcut to succeeding in a math class. Students of math typically spend a lot of time learning all they are expected to learn. As most things worth doing require a lot of effort, and most things we learn take a lot of practice to get right.

If you want to learn how to play a musical instrument, you will have to practice for hours playing simple tunes before you can play a more complex melody.

If you want to master any sport, you will have to practice a lot in order to develop the coordination and endurance that you need.

Contents

Josiah Gilbert Holland got it right when he said, *"There is no great achievement that is not the result of patient working and waiting."*

There is no secret formula to succeeding in math. But there are a lot of things you can do that will substantially improve your chances of success, and I have made it my mission to share a great many things in this book that helped me as a graduate student and helped many of my students as well.

Most students who succeed in math, practice some or all of the strategies given in this book. I also believe applying these strategies had a huge hand in my own experience earning A grades in all of my graduate level math classes. These strategies are applicable to you no matter what math class you are taking–whether you are in high school, college, graduate school, or you have not taken a math class in many years. These are the pathways to finding your way through the maze of your next math class.

CHAPTER 1 - PLANNING AHEAD

CHAPTER 1. PLANNING AHEAD

> ◈ Pathways
>
> - Who is the instructor?
> - What is the teaching style?
> - What resources will be required?
> - Where will you get support?

Planning ahead is a critical first step toward succeeding in a math class. Planning ahead involves researching the class and the instructor before the first day. You should learn all you can about the teaching style and expectations of the class.

♥ Who is the instructor?

Consider emailing the instructor before the first day of class to ask them about their teaching style, course materials, and recommended supplemental resources. Also consider asking other students who have taken the instructor's class.

See what you can find out online, as well. Some instructors will have a permanent class website with materials for their course. Try looking them up on the college/university site, or doing a web search for their name and the class.

Fill in the blanks with responses for your course:

> Instructor: _____
>
> Email: _____
>
> Website: _____

♀ What is the teaching style?

Does the instructor teach with a traditional style or take a more active learning approach? Does the instructor use a textbook, videos, or both? Is there an exam every week, every few weeks or only at the end of the term?

If the teaching style is traditional

The class may be a lecture from the instructor each day. If there is a lecture every day and corresponding course materials, review those materials before going to class each day. This will make it easier to follow along with the instructor's presentation and ask questions.

CHAPTER 1. PLANNING AHEAD

If the teaching style takes a more active learning or student centered approach

The instructor may have you work in groups with other students or alone on assignments during the class. If there are corresponding lessons for each day, make sure that you review those lessons before going to the class.

Fill in the blanks with responses for your course:

Teaching style: _____

Frequency of Exams: _____

♥ What resources will be required?

Will there be a textbook that you are expected to purchase before the first day? What type of calculator is recommended and what resources are recommended?

If you are taking the course online or with an online component, you may be required to purchase an access code and have access to special equipment for taking exams. Find out what online learning resource, if any, the instructor will use.

If a textbook is mandatory, also look for supplementary resources like a solutions manual, a study guide, and online resources.

Does the instructor require graphing software or some other software for analyzing data, such as in a calculus or statistics class? If such software or a calculator is required, you should acquire and become familiar with it before the first day of class.

> "Look for supplementary resources like a solutions manual, a study guide, and online resources."

List items required for your math class:

1. _____
2. _____
3. _____
4. _____

9 What if there is no textbook.

With all the technology that is available today, some teachers may not use an actual print textbook. They

CHAPTER 1. PLANNING AHEAD

may have a video text, a series of videos that function as lectures and a textbook, an electronic text only, or they may just have example videos and no text at all.

If there is no print or electronic textbook.

Consider asking the instructor for supplemental recommendations or a list of the course topics. You can also check the course syllabus for a list of topics that will be covered in the course. Then look for a book with good reviews that also covers those topics.

Note: If there is a video text provided by the instructor you should watch those videos whether you have a supplemental text or not. Just to be sure that you do not miss something important that is only explained in the videos.

If there is only an electronic textbook.

Consider ordering a print copy yourself by sending the electronic file to a printing company near you. Most printing companies will print a large file for a reasonable price and give you multiple options for paper quality and binding.

♀ Where will you get support?

What resources are at your disposal for getting support with the class? Is there a math lab or tutor option? Are their computers and internet that you can use on campus? Are there other resources for asking questions and getting help?

You should investigate all these things before the first day of class. You should have all the things that need to be purchased, before the first day of class and get familiar with the topics that will be covered in the course.

List options for getting math support:

1. _____
2. _____
3. _____
4. _____

CHAPTER 1. PLANNING AHEAD

Questions

1. What does it mean to plan ahead?

CHAPTER 2 - OPENING YOUR MIND

CHAPTER 2. OPENING YOUR MIND

> ✖ Pathways
> ─────────────────────────────
> - Take an active approach.
> - Be open to change.
> - Avoid the anxiety bug.

No matter your past experiences with mathematics, you must keep an open mind, knowing that if you take an active approach to your learning you CAN succeed.

♀ Take an active approach.

Active learners plan ahead. They pay close attention to the expectations of a course, and set time aside every day to spend on the course. They take note of important due dates, attend classes regularly, and keep track of all course announcements.

16 Things Active Learners Do

1. Research class before the first day.
2. Purchase required materials before the first day.
3. Overview class materials before the first day.
4. Contact the instructor about concerns..

5. Read the course syllabus on the first day.

6. Take note of important dates.

7. Make a study plan.

8. Review each class lesson before class.

9. Attend every class.

10. Take notes in class.

11. Ask questions when they do not understand.

12. Ask questions when they want to know more.

13. Participate in class discussions.

14. Complete all graded assignments.

15. Review for exams.

16. Clarify mistakes and learn from them.

Be open to change.

There is power in attitude. Our attitude affects how we perceive things, how much we learn, and how much we retain what we learn. Consider your own attitude as you begin your math class. Are you open to learning math? Do you believe you can learn math? Do you dislike math?

CHAPTER 2. OPENING YOUR MIND

Are you open to learning math?

Why are you taking a math class? How will taking a math class help you attain future goals? Consider why you are taking a math class, and keep your reason in mind, so that you can easily recall it if you struggle in the class.

Do you believe you can learn math?

Even if you struggled with math in the past, it does not mean you cannot learn math now. If you are taking the class with an open mind, for the right reasons, and you are willing to apply yourself to learning the content.... you should succeed. Do not let any preconceived ideas about math hinder your success.

Do you dislike math?

You do not have to like math to learn math, but it is easier to learn and retain something if we can find something to like about it. Consider how the math you will be learning applies to your future goals. If you need specific examples, consider doing an online search of math in your field or ask your instructor for some examples. Then know that, by learning some math, you are gaining problem solving skills that many employers consider very valuable.

It may also help you to know that recent research in Neuropsychology has shown the mind is a muscle that we can grow if we have the right mindset. People with a growth mindset believe they can grow their intelligence with effort and time. Take a growth mindset toward math and grow your problem solving skills.

Avoid the anxiety bug.

If you believe that you are somehow incapable of succeeding in a math class, then you just might not succeed. Sometimes when we think and expect something to be hard, in our minds, we make it hard; and we build walls in our mind that hinder our learning. Ask yourself if you have any pre-conceived notions about math that have you believing that you are not a math person. Forget those notions.

>"Sometimes when we expect something to be hard we make it hard, and build walls in our mind that hinder our learning."

CHAPTER 2. OPENING YOUR MIND

Math is a language and to become fluent in any language takes time. You have to begin with the basics and build up your knowledge one concept at a time.

To get good at anything requires a lot of hard work and practice. Resolve to work hard at solving problems and know that, if you keep practicing, you will succeed.

Questions

1. What does it mean to open your mind?

2. How will you open your mind?

CHAPTER 3 - STUDYING SMART

CHAPTER 3. STUDYING SMART

> ◇ Pathways
>
> - Where are you studying?
> - What do you need when you study?
> - How are you studying?
> - How can you stay motivated?

There is an Aesop fable about a hare and tortoise which teaches us that *steady as it goes wins the race.* This principle can also be applied to math, because **consistency** with studying is a key component to surviving any math class.

Set a goal to spend time on your math class each day. *Think about how to best utilize the time so that you are studying smart instead of studying hard.* If you are spending a lot of time on math but not solving any problems, then you are studying hard and not studying smart.

As you study, *you should be accomplishing things that bring you closer to completing your learning outcomes.* If you are not doing this, then you need to review where you are studying, how you are studying, and what you are studying.

> "If you are spending a lot of time on math but not solving any problems, then you are studying hard and not studying smart."

♀ Where are you studying?

Study in an environment that helps you focus. Some people focus better in complete silence. Some people focus better when there is a lot of white noise, and some people focus better while listening to music. Think about what type of environment helps you focus and choose a place that will provide this environment while you study. You will accomplish more if you can focus and have few interruptions.

If you think you will have lots of questions, then consider where you may study and ask your questions, such as in a classroom, a learning center, or even near someone you know personally who can help you with the material.

CHAPTER 3. STUDYING SMART

List ideal locations for studying math:

1. _____

2. _____

3. _____

4. _____

♥ What do you need when you study?

Do you have sufficient resources for your studying? Do you have a textbook and supplemental resources to help you understand the concepts? If applicable, do you have a reliable internet connection and a reliable computer to use for online assignments?

Do you have a sufficient calculator or math software for computation? For an algebra class you should have a multi-view scientific calculator and for a calculus class you should have a graphing calculator or access to graphing software and know how to use them. Desmos and Geogebra are two examples of online calculator software that are free to use.

You should also have a notebook or scratch paper for writing down your work while solving problems. Having your work written down will give you something to review with later.

List what you need when you study:

1. _____
2. _____
3. _____
4. _____

? How are you studying?

Do not set time goals for studying, such as, one hour here and two hours there. Set goals that will bring you closer to learning the expected outcomes of the course. For example, think about how many problems you need to solve when you study. Consider how many are in the assignment and divide them over the time that is available. Specifically, if you have twenty problems to solve in two days with two hours to study each day, then you will want to solve five problems each hour or at least ten problems in each study session.

$$\frac{\text{Problems to solve}}{\text{Days until due}} = \text{Problems per day}$$

CHAPTER 3. STUDYING SMART

How can you stay motivated?

What strategies can you apply to further help you focus and accomplish your learning outcomes?

Avoid multi-tasking such as watching tv while studying; and consider listening to music that will help you concentrate and clear your mind.

Consider taking breaks to refresh your mind, and give your mind time to process concepts and problems.

Give yourself small rewards for making progress toward your learning goals. This will help you maintain your motivation for studying.

> "It is not how much time you spend studying, but how you use the time that you have."

Though it may not be an option, studying when you are most alert is ideal, and staying up all night should not be necessary if you are making good use of your study time. It is not about how much time you spend studying, but how you use the time that you have.

Questions

1. What does it mean to study smart?

2. What are three things you can do to help you study smart?

CHAPTER 4 - USING A TEXTBOOK

CHAPTER 4. USING A TEXTBOOK

> ✂ Pathways
>
> - Using the definitions.
> - Using the examples.
> - Using the exercises.
> - Being an active reader.

Most math textbooks are laid out the same way with definitions, examples, and exercises.

♀ Using the definitions.

Definitions are the keys to understanding the "why's" in what you are learning. They can be a few words or a set of rules for you to follow, such as, the order of operations for simplifying an expression.

> Order of Operations for Simplifying an expression
>
> 1. Grouping symbols, beginning with the innermost grouping symbols
> 2. Exponents
> 3. Multiplication and division, from left to right
> 4. Addition and subtraction, from left to right

Definitions are written purposely. Pay close attention to every word in a definition as missing part of the definition may make it hard to solve corresponding problems.

For example, in the order of operations it says to multiply and divide *from left to right.* In the following example with multiplication and division, notice that we get different results if we simplify from the left versus from the right.

Example 4.1 Simplify.

$$12 \div 4 \times 3$$

Solution:

Simplifying **left to right** by dividing 4 into 12 then multiplying by 3 leads to a result of 9.

$$12 \div 4 \times 3$$
$$3 \times 3$$
$$9$$

Simplifying **right to left** by multiplying 4 and 3 to get 12 then dividing 12 into 12 leads to a result of 1.

CHAPTER 4. USING A TEXTBOOK

$$12 \div 4 \times 3$$
$$12 \div 12$$
$$1$$

The correct process for simplifying an expression with multiplication and division is to simplify from *left to right*. If you did not notice this clarification in the definition then you may have simplified the above example from *right to left* and ended up with the wrong solution.

If you are having trouble solving a problem in a given section, try reviewing the definition for key information that you missed.

Definitions should be your primary source of clarity in a given textbook section and you should review them again and again until you have mastered the section.

♥ Using the examples.

Examples are a means of seeing how to apply a definition to relevant problems, and they should be used along with definitions to master each concept. The following example uses the order of operations to simplify an expression. Notice how each rule is applied from the definition given in the previous section.

Example 4.2 Simplify.

$$9^2 \div 3(5-3) + 10$$

Solution:

Simplify in **parentheses** first, $5 - 3 = 2$

$$9^2 \div 3(2) + 10$$

Simplify the **exponent**, $9^2 = 81$

$$81 \div 3 \times 2 + 10$$

Multiply and divide from left to right

$$81 \div 3 \times 2 + 10$$
$$27 \times 2 + 10$$
$$54 + 10$$

Add

$$64$$

Make sure you are using definitions and examples to understand concepts. You should understand which definitions are being applied to each example and how they are being applied. If examples are your only means for understanding concepts, then you may be missing things along the way and eventually the holes

CHAPTER 4. USING A TEXTBOOK

in your learning will make future more advanced concepts difficult to understand.

Examples are also a great resource for when you get stuck on a problem. If you get stuck on a problem, look in the section for an example that is similar to the problem. Close review of a similar example may help you see how to get unstuck.

♥ Using the exercises.

Exercises will generally cover the main definitions from the section in the order they were presented. Following are a few different kinds of exercises you may see in a math textbook.

Some exercises will require you to apply a definition in unconventional ways. Like finding errors in a solution to an order of operations problem or determining the correct expression for an order of operations problem.

Some exercises may combine a current definition with a previously learned definition from a previous section. Like simplifying an expressions with order of operations rules that also contains integers or fractions.

Some of the exercises may also require you to stretch your mind and think outside the box. Like a word problem that has to be solved in a specific order using the order of operation rules.

You should have enough information in the current and previous sections of your textbook to solve even these exercises if you are clever enough, patient enough, and precise in your attack.

♀ Being an active reader.

We read most books by passively reading line by line; however, textbooks are different. To get more out of your textbooks, you should be reading them actively and interactively. When you are given a reading assignment from a textbook, you should start by surveying the main ideas and key terms.

- How does the passage you are reading fit into the entire chapter and the entire book?

- What is the context of the reading?

- Why are you reading it?

- How does it fit in with what you read previously, or what you will be reading next?

CHAPTER 4. USING A TEXTBOOK

Do not be afraid to highlight, underline, and circle key words and phrases in your textbook. Consider also writing words in the margins.

If you do not understand something about the reading, seek clarification.

15 Things Active Readers Do

1. Do the assigned reading ahead of time.
2. Look over the assigned reading before reading.
3. Pay close attention to headings and emphasized text.
4. Pay close attention to definitions and examples.
5. Consider how the current reading fits into what they already know.
6. Write notes in the margins.
7. Underline, highlight, and make note of important information.
8. Determine key ideas in the section.
9. Write down questions about what they read.
10. Quiz themselves on what they read.
11. Seek clarification as necessary.
12. Read more than once, if necessary.

13. Read aloud.

14. Look for the main ideas in each paragraph.

15. Consider the overall purpose of the reading.

Questions

1. What happens if examples are your only means for learning concepts?

2. How can you be an active reader?

CHAPTER 5 - LEARNING THE BIG IDEAS

CHAPTER 5. LEARNING THE BIG IDEAS

> ❖ Pathways
> -
> - Minimizing necessary memorization.
> - Maximizing retention.
> - Making mind maps.

There is a big difference between doing math and understanding math. Being able to do math is seeing the trees, but understanding the concepts (the definitions) is seeing the whole forest, the big picture.

A big idea in math is a general rule that can be applied to every problem under a specific context. For example, the general rule for combining like terms is to *combine the number parts and keep the common variable part*. If you do not know this general rule for combining like terms, then it is unlikely that you will correctly combine your terms <u>every time</u> you simplify an expression.

♀ Minimizing necessary memorization

If you can see the whole forest, then you can look for patterns in the trees so that you do not have to memorize the look of every single tree. Similarly, learning definitions in mathematics helps you see patterns in solving problems, so that you do not have to memorize every single problem.

Therefore, you will have less to memorize for an exam, if you take time to reflect on the main ideas (definitions) given in each chapter. You need to know what they are and how to apply them with only slight variations. This is really the trick to doing well on a math exam.

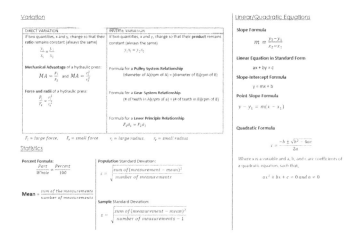

Figure 5.1: Technical Math Notes

One way to reflect on main ideas is to make a sheet of notes containing all of the main ideas from the current chapter(s). If possible, also include an example with each main idea, showing how to apply it. Figure 5.1 is a sheet of notes with main ideas for a technical math exam. Some teachers will let you use a sheet of notes like this when you take a test. Even if your teacher

CHAPTER 5. LEARNING THE BIG IDEAS

does not allow notes, making a sheet of notes like this will help you organize all that you learned, and give you less to memorize.

♀ Maximizing retention.

Learning the main ideas will also help you retain and recall what you learned. The better you understand something the more likely you will be able to recall it in the future. Understanding the big picture, is needing only to learn three types of trees to know the whole forest rather than memorizing every tree. This is the trick to maximizing retention, placing complex ideas in more simple terms that are easier to memorize.

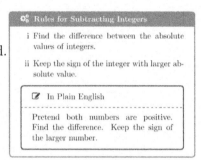

Figure 5.2: Rules for Subtracting Integers

Figure 5.2, is an example of using simplified phrasing for easier memorization. The two rules given first, represent a complex explanation for subtracting integers. The "plain English" given second places those rules into more simple terms that are easier to memorize. As you learn mathematics, consider how you may think about the key ideas in simple terms to improve your retention.

9 Making mind maps.

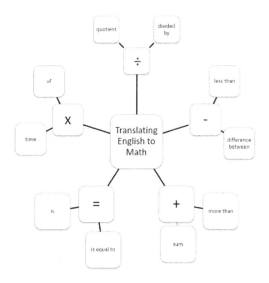

Figure 5.3: Translating English into Math Mind Map

Another great strategy for learning big ideas in math is by using mind maps to organize key concepts and their supporting details. Figure 5.3 is a demonstration of a mind map for "Translating English into Math". The center of the mind map is the big idea, Translating English into math. The supporting details, basic operations (+, - , x, ÷), stem from that big idea, and the next branches show the English words

CHAPTER 5. LEARNING THE BIG IDEAS

for those operations.

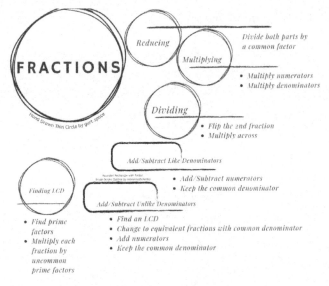

Figure 5.4: Fraction Mind Map

Figure 5.4 is another example of a mind map where Fractions is the big idea inside the largest shape. The smaller shapes contain rules for simplifying fractions, and the list under each rule shows the steps for applying each rule.

Questions

1. What are "Big Ideas" in math?

2. Why should you learn the "Big Ideas" in math?

CHAPTER 6 - SHOWING YOUR WORK

CHAPTER 6. SHOWING YOUR WORK

> ✜ Pathways
> ----
> - Show work to save time.
> - Show work to get help.
> - Show neat work.
> - Show work for test success.

Showing working in math is writing down the steps you follow when solving a problem. Showing this kind of work can help you save time. It can help you communicate your problem solving process. It can help you find errors in your problem solving process, and it can help you prepare for exams.

♀ Show work to save time.

Writing things down takes time. It may seem like doing work in your head would save you time, but if you make a mistake you will have to begin again and you will ultimately lose time starting over.

> "It may seem like you are saving time by doing work in your head, but if you make a mistake you will have to begin again and lose time starting over."

If you are writing things down as you go, and your answer is wrong, you can simply check your work for mistakes and save time not having to begin again.

⚑ Show work to get help.

Suppose you get stuck on a problem or you are not sure where to go next or why your answer is incorrect. If you wrote down your work, your instructor may review your work and determine specifically what you missed or what you do not understand.

If you are not showing your work, then your instructor may not know what part of the process you do not understand.

⚑ Show neat work.

Do not just show work. Also pay close attention to how work is shown.

CHAPTER 6. SHOWING YOUR WORK

Since math is a language, there is a generally accepted vocabulary and structure of phrases necessary for clear communication. Learning the generally accepted ways for showing your work in math will help you communicate to your instructor and others what you know.

Consider Example 6.1.

Example 6.1 Simplify.

$$20 \div 4 \times 5 + 2 - 7$$

Solution:

$$20 \div 4 \times 5 + 2 - 7$$
$$5 \times 5 + 2 - 7$$
$$25 + 2 - 7$$
$$27 - 7$$
$$20$$

As demonstrated with Example 6.1, neat work is shown line by line with each simplified result written underneath the previous expression. This makes it easier to follow your work.

Compare the work shown in Example 6.1 to the work shown in Example 6.2.

Example 6.2 Simplify.

$$20 \div 4 \times 5 + 2 - 7$$

Solution:

$$20 \div 4 \times 5 + 2 - 7$$

$$20 \div 4 = 5 \qquad 5 \times 5 = 25$$

$$25 + 2 = 27 \qquad 27 - 7 = 20$$

Do you think it is easier to follow the work in Example 6.1 or Example 6.2?

Consider the work you are showing when you complete each assignment. Showing neat work will make it easier for others to understand what you did. Showing neat work will also give you something to review with later when you are preparing for a test.

♥ Show work for test success.

Showing your work will help you prepare for a test. Research has shown that we gain a better understanding of what we are learning when we take hand written notes. Writing things down can also help you retain them for later.

When you take a test you should know the processes

CHAPTER 6. SHOWING YOUR WORK

for solving the problems so well that you can do the work without thinking very much. This will help when you are under the pressure of time. Writing down problem solving processes over and over while doing homework, will help you memorize them for a test.

> "When you take a test you should know the processes for solving problems SO WELL that you can do the work without thinking."

Questions

1. What are *three* reasons you should show work in math?

CHAPTER 7 - SOLVING PROBLEMS

CHAPTER 7. SOLVING PROBLEMS

> ◈ Pathways
> -
> - Solving word problems
> - Finding careless math errors
> - Help! I am stuck!

Some problems in mathematics are straight forward and require only simple computation to find a solution. Then other problems are layers of complex ideas that must be cleverly unraveled one step at a time before a solution can be discovered.

♀ Solving Word Problems

When solving a complex word problem, begin by writing down information about the problem. You must understand the problem before you can solve it.

What is the question?

First determine the specific question you are being asked to answer. You cannot answer a question unless you clearly understand what is being asked.

What information are you given?

Is there any information given that you do not need? Sometimes information is thrown into the problem that you do not need, to make the problem more challenging help you sharpen your critical thinking skills.

Is there any information not given that you do need? For example, a problem that requires you to find the volume of a cylinder may not actually in clude the formula for finding the volume of a cylinder, because you are expected to know it already.

In a real life problem you would have to separate the information you need from the information you do not need, and know where to look for the tools that will help you solve the problem.

Can you work backwards?

Try working backward from the solution. What is the previous thing you would need to know to answer the question and what would you need to know before that and before that and before that until you reach the information given to start the problem.

CHAPTER 7. SOLVING PROBLEMS

Can you organize your steps?

Use your backward solution to determine the proper steps you must follow to answer the question and solve the problem.

9 Finding Careless Math Errors

Many wrong answers in math are a result of careless errors. If you did not get the right solution, try looking for one of the following common errors:

Common Math Errors

- **Copying Error** This happens when we copy a problem and our copy does not match the original problem.

- **Transposing Error** This happens when we write the digits of a number in reverse order.

- **Sign Error** This happens when we drop or add a negative sign in our computation.

- **Legibility Error** This happens when we cannot read our own handwriting.

- **Directions Error** This happens when we misread directions.

- **Computational Error** This happens when we add instead of subtracting, multiply instead of dividing, etc.

- **Number Error** This happens when we write a different number from what we are suppose to write.

- **Calculator Error** This happens when we put the wrong information in the calculator.

Do you find yourself making certain errors from this list more frequently than others? Consider the errors you make most frequently and what steps you can take to avoid making those errors in the future. This will help you get the correct solution immediately more often, and it will lead to less errors on a test as well.

Help! I am stuck!

Everyone gets stuck on a problem at some point, and you may need to ask for help; but, before you do, there are a few things you can try.

Check your work

First check to see if you copied down the question correctly. Then also check for errors in your reasoning. Does each step logically follow from the previous step to your solution?

CHAPTER 7. SOLVING PROBLEMS

Check the Definition

What is the big idea behind the problem? Is there a general definition or set of rules you are suppose to follow? You may need to review them and reaffirm that you are not missing any part of the required steps or definition.

Check yourself out for a break

Sometimes it helps to leave a problem when you are stuck and do something other than math for a while. This will give your mind time to process the problem and discover new ways of thinking about it. Then, when you return to the problem, you will also be viewing it with fresh eyes that may see something new that your tired eyes could not see before.

Check with a different pair of eyes

Have someone review your work. Sometimes a different pair of eyes will see something your eyes may have missed. Sometimes, when we know what something is suppose to say, we see what we know it is suppose to say, instead of what is really there.

Ask for Help

Once you have made a decent effort on the problem, it is better to ask for help and receive clarity rather than to never understand the problem. Not understanding homework problems may lead to holes in your knowledge and poor test scores. Remember, math builds on itself.

Questions

1. What math errors do you make most frequently when solving a problem?

CHAPTER 7. SOLVING PROBLEMS

2. What steps can you take to avoid making those errors in future problem solving?

CHAPTER 8 - ASKING FOR HELP

CHAPTER 8. ASKING FOR HELP

> ✜ Pathways
> ---
> - Before getting help.
> - While getting help.
> - After getting help.

Timid and Determined are taking a math class together and studying for a test when they come across a problem they do not know how to solve.

Timid does not want to ask for help and he tells himself that it is only one problem that will probably not be on the test. However, the next day THE problem IS on the exam and worth a large number of points. Because Timid did not ask for help, he is not able to solve the problem and he does not receive an A on the exam.

Determined really wants to get an A on the exam so Determined decides to seek help and learns how to solve the problem. When Determined sees the same problem on the test the next day, he knows how to solve it and he gets an A on the test.

Do not be afraid to ask for help. Not getting help when you need help can have a huge affect on your success.

♀ Before getting help

Review the "Problem Solving" lesson of this book. Specifically, the section titled, "Help! I am stuck!"; and follow the recommended steps given under this section. The more questions you can answer on your own, the less time you will spend being confused. Take time to review corresponding definitions and examples before asking another person for help.

However, do not spend a lot of time going in circles. If you do not know the answer, after trying a couple things on your own, ask for help.

Also beware of getting a correct answer without knowing why you answer is correct. You should be able to justify your thinking. If you cannot justify your thinking, then you do not truly understand the problem, and you should ask for help.

CHAPTER 8. ASKING FOR HELP

♥ While getting help

Ask specific questions; and know ahead of time what specific questions you want to ask. A general question is saying "I do not understand this chapter". A more specific question is, "Can you show me how to add fractions with different denominators?"

If you are asking about a specific problem from a textbook or homework assignment, be sure to reference it with where it is located in the textbook and/or homework assignment. This will give your helper context for your question, and make it easier for them to give a specific answer.

♥ After getting help

Take some time to reflect on your experience. Did you get the help you needed? Should you do anything different next time?

Questions

1. What are some reasons you might not want to ask for help?

2. What are some reasons you should ask for help?

CHAPTER 9 - TESTING STRATEGIES

CHAPTER 9. TESTING STRATEGIES

> ◈ Pathways
> -
> - Before taking a test
> - While taking a test
> - After taking a test
> - When you have test anxiety

Tests are meant to show what you know. If this is how a test is being utilized, then you should do well on a test if you prepare properly for the test.

♀ Before taking a test

Before you take a test, you should investigate what the test will be like, make sure you know the main ideas covered by the test, take a practice test, seek necessary clarifications, and more. Even if you think you understood the material when you were completing the homework, reviewing it will help you remember it for your test.

Investigate the test

What material will the test cover? How many questions will be on the test? What types of questions will be on the test (e.g. multiple choice, short answer, etc.)? How much time will you be given for the test? Will you be allowed to use a calculator? Will you be allowed to have any notes and/or scratch paper?

List things you need to know about your next test:

1. _____

2. _____

3. _____

4. _____

Main ideas and supporting details

As you are reviewing for your test, look for the main ideas and their supporting details. A main idea is a concept. The supporting details are the different applications for the concept.

For example, "Rules for Solving Equations" is a main idea. The supporting details would be how to apply the rules to different types of equations: equations with variables on both sides, equations with fractions, equations with decimals, etc.

CHAPTER 9. TESTING STRATEGIES

Make a cheat sheet

Even if you are not allowed to use notes when taking a test, you should still make a sheet of notes as though you are going to use it for the test. Your sheet of notes should contain all the main ideas that you should know for the test. Making this cheat sheet of notes will help you organize all of the concepts and feel less overwhelmed. Remember, it is much easier to look at the whole forest and see that there are only three different types of trees instead of trying to remember every single tree.

Make a practice test

Some instructors will provide a practice test that you can use to prepare for the actual test.

If your instructor does not provide a practice test, then you should make your test based on what the instructor says you should review for the test.

Then take the practice test under the expected test settings, with the same time limit and permitted resources.

Ask questions

If there is anything you do not understand well before taking a test, you should always seek clarification before the test. For a demonstration of what might happen if you do not seek clarification before taking the test, see the Timid and Determination scenario given at the beginning of the "Asking for Help" chapter. Review the "Asking for Help" lesson if you need tips for seeking help.

Stress-Relief Techniques

Consider practicing deep breathing techniques before taking the test, that you may use during your test if you are feeling stressed. Deep breathing elevates the amount of oxygen traveling to your brain, helping your brain to relax.

Rest

Remember to get plenty of rest and eat well before taking a test. It is hard to think clearly and do our best if you we are hungry or tired. Consider how you can prioritize tasks so that you get the rest and care that you need.

CHAPTER 9. TESTING STRATEGIES

♥ While taking a test

Know what resources you are expected to have for your test and go to the test prepared with those resources. For example, maybe you need a calculator or scratch paper when you take the test.

If you are not allowed a sheet notes, consider taking a few minutes at the beginning of the test to write down all the main ideas you can remember from the relevant chapter(s).

Look over the entire test before you begin solving any of the problems.

Start with the problems that you know immediately how to solve and save the ones that will take more time for last. It will be easier to focus on the harder questions if you know you answered everything else already.

If you get stuck on a question, skip it and save it for last. Maybe an idea for answering the question will occur to you while you are working on the other problems.

Always show work whenever possible. The potential for partial credit, if your instructor gives partial credit, can be the difference between passing and failing a test.

If you are not sure how to answer a question, consider what you do know and write it down. Maybe you will think of something else you know then something else and figure the problem out after all. If you do not figure out the entire problem, you might get partial credit for what you do know.

Try not to panic while taking a test. If you feel anxiety creeping in, take a few deeps breaths—breath in through your nose counting to five and filling your belly with air then breath out counting to five. It is a proven means of lowering stress in your body and it will work if you do it correctly.

If you have time left over when you are finished, then check your work before turning in your test. Check for the *common math errors* given in the problem solving section of this book (copying, transposing, computational, calculator, etc.).

♥ After taking a test

You have finished your test. Now it is time to relax. If possible, take a break from studying for the rest of the day.

When you get a chance to see your test results, be humble if you did better than you expected and resilient if you did worse than you expected.

CHAPTER 9. TESTING STRATEGIES

If you did better than you expected.

Well done! Reward yourself with something you enjoy doing. Then take some time to reflect on what you can do going forward to ensure that you continue to do well.

If you did worse than you expected.

Do not give up. Analyze your test and consider what went wrong, and what you can do differently going forward to ensure a better result next time.

Following is a list of the *most common test taking errors*:

- **Following Directions** Misreading or misunderstanding directions

- **Computation** Knowing how to solve a problem, but making simple mistakes that lead to an incorrect solution.

- **Concept** Not understanding how to solve a problem.

- **Application** Knowing the concept, but not knowing how to apply it to a specific situation.

- **Testing** Making mistakes in how you approach a test.

- Spending too much time on one or more problems.
- Not finishing one or more problems.
- Going too fast.
- Second guessing answers.
- etc.

- **Studying** Studying the wrong material or spending too much or too little time on test preparation.

What were your most common errors and how can you avoid them on a future test? What are some specific things you can do before your next test to ensure better results?

♥ When you have test anxiety

Most test anxiety is due to either feeling under-prepared or expecting the test to not go well.

If you are not sure how to prepare for a test, see the suggestions given in the following sections of this lesson.

If your mind wants to dwell on the worst than can happen then consider the worst. What will you do if the worst thing does happen? What will need to happen next, and how will you deal with it moving

CHAPTER 9. TESTING STRATEGIES

forward. Sometimes having a plan for the worst result can help remove the pressure that it may cause, and allow you to focus on hoping for the best result.

Maybe you will get a poor grade?

Grades are not everything, and sometimes we can get so focused on getting a good grade, we miss truly understanding and retaining what we are learning. Getting a good grade should be the preferred outcome, but grades do not have to determine your happiness or your success.

Maybe you will have to take the class again?

Taking the class over again is a good idea if you did not learn much of the material. The class may also be easier the second time you take it. Taking it again may mean taking longer to finish your program; but, keep in mind, that a few months is a drop in the bucket compared to your entire life. It seems like a long time now, but five years down the road it will not seem long at all.

Whatever the reason for your concern, find a way to look at it in a positive light so that you will not feel anxious about it. Find a way to turn your lemons into lemonade. Do not let your emotions decide the outcome of your test.

Questions

1. What are your most common errors when testing?

2. What are at least *three* things you can do to ensure continued or future success in testing?

CHAPTER 10 - LEARNING REMOTELY

CHAPTER 10. LEARNING REMOTELY

> ✖ Pathways
>
> - How you should participate.
> - How to complete assignments.
> - How you should study.
> - How to take online exams.

Technology has made it possible to form many parallels between learning math in a classroom and learning math remotely. With these options, more and more instructors are embracing remote learning. We have electronic textbooks, classroom discussions in discussions forums, online calculators for calculations, and, depending on the instructor, we may even have lectures and group work in a video conference environment with interaction taking place in real time.

However, there are still some differences in how you should participate, complete assignments, take exams, and study when you are taking a math class remotely.

♀ How you should participate.

In most online classes, you will interact in real time with the instructor only on rare occasions or not at all. This means you will need to sign-in to the class on a regular basis to participate.

Some instructors communicate frequently through email, announcements and discussions posts. You will want to keep track of all instructor communication for important due dates and changes to the course.

Some instructors do not communicate frequently and it may even take several days to get a response to a question or request. It may also be the case that the instructor's response is not helpful. Fortunately for you, there are an abundance of resources available to help you succeed in a math class. If the resources provided by the instructor are not right for you, try finding resources that will work for you. Consider purchasing supplemental materials, subscribing to relevant video channels on Youtube, using online tutoring services, or consider who you can reach out to for help and suggestions.

CHAPTER 10. LEARNING REMOTELY

List resources that are right for you:

1. _____
2. _____
3. _____
4. _____

If you are not getting the answers you need, you should also consider how you are communicating your questions. Are you being specific in your requests? For example, if you are asking about a specific problem, you should include the name of the assignment and the problem number. The more specific your questions, the easier it will be for the instructor to give you the information you need.

♀ How to complete assignments.

Some instructors will have you complete problem assignments on paper then scan and submit them electronically. If you do not already own a scanner, there are many different scanning apps that can be downloaded to a smart phone. Most of them are free and easy to use.

List scanner options:

```
1. _____

2. _____

3. _____
```

Other instructors may have electronic assignments where you can submit your answer to each question online. If they do not require you to show work, consider keeping a notebook of your work anyway. It will help you memorize problem solving processes and give you something to review with for assessments. If you prefer to have a paper assignment, check with your instructor to see if you can print out the electronic assignments. Then you can write your work on the printed paper before submitting your answers online. This is also useful if you have a slow internet connection, or you do not have regular access to a computer.

◉ How you should study.

Consistency is key when studying for a remote class. You may not be able to get help as easily as you could in a face-to-face class, so it is even more critical that you do not fall behind.

Look for a study guide or assignment schedule and use them to make smart goals for studying that are

CHAPTER 10. LEARNING REMOTELY

specific and manageable.

Try to complete assignments one or two days before they are due in case something unexpected comes up.

Then if you need more tips for studying, also see the "Studying Smart" chapter of this book.

♥ How to take online exams.

Some instructors will have you complete your exam in a testing center just, like you would in the classroom, with a printed test and the ability to show all of your work. While other instructors may require you to take your exam online with a proctoring company like Honorlock or Proctor U.

Taking an exam in a testing center

Taking an exam in a testing center is not much different from taking an exam in the classroom. Just make sure you have everything you need before you go to the test, because the instructor will not be right there to help you. Make sure you ask any questions you have before going to take the test. Make sure you have appropriate identification with you to verify who you are. You should also have scratch paper, extra pencils, and extra batteries for your calculator.

Taking an exam online with a virtual proctor.

The instructor should provide information to review about the virtual proctor. Make sure you review this information thoroughly before taking the exam. Here is a list of things to consider if you are taking an exam online.

- Take the exam in a place with minimal distractions.

- Make sure you have a fast and reliable internet connection.

- Make sure you have a fast and reliable computer.

- Make sure you have a working webcam.

- Practice using the proctoring service before taking the exam, if possible.

- Have photo ID for verifying your identity.

- Review all instructions for the exam before taking the exam.

- Have all tools you may need within reach, when you take the exam.

- Remember that everything you do and say while taking the exam is being recorded.

- Have a back up plan in place in case you lose access to the resources you need.

CHAPTER 10. LEARNING REMOTELY

- Communicate with your instructor if you have any technical difficulties that hinder your test results.

Depending on your instructor, you may not be able to show work for an online test, so practice a lot before hand and try to minimize the errors you are making when solving problems. If you have time, while taking a test, review your work and answers before submitting your test.

Questions

1. Why should online assignments be submitted 1-2 days before they are due?

2. What can you do if your computer breaks down the day before you are planning to take an online exam?

CHAPTER 11 - AVOIDING BURNOUT

CHAPTER 11. AVOIDING BURNOUT

> ◈ Pathways
> -
> - Taking math with lighter classes.
> - Not procrastinating in math.
> - Not making math too hard.

The best way to avoid burnout when taking math is to take math when your schedule is open and flexible, so you can spend as much time as is necessary to succeed in the class. However, this may not be an option for you, and it is still possible to succeed if you study smart to get the most out of your study time, and you avoid the pitfalls mentioned in this chapter.

◉ Taking math with lighter classes.

Not all classes require the same amount of effort, some class are light classes that require less effort while other classes are heavy classes that require more effort. A light class does not require much homework or study time. A heavy class requires a lot of time studying to do well. Most students must spend a lot of time studying *consistently* in a math class to do well. With this is mind you should not take math with other heavy classes that also require a lot of time studying. If you must take other classes while taking math, then consider taking classes that will not require as much effort as a heavy class. This will allow you to contribute more time to math and improve your chances of succeeding.

♀ Not procrastinating in math.

In some classes you may be able to skip or miss some material and still understand other material. This is not the case in a math class. In a math class, concepts build on each other, so if you miss some of the material you may struggle with other material.

> "Most students who fall behind in math do not succeed in math."

If you procrastinate and fall behind in your math class, you will have to double your efforts to catch up, learning fundamental concepts while also learning more advanced concepts. This can get overwhelming very quickly. Most students who fall behind in math do not succeed in math, so make every effort to stay on track by being consistent with your study habits and your approach.

♀ Not making math too hard

It is a common pitfall in math to make math problems or concepts harder than they really are when we expect them to be difficult. Learning new things can seem hard at first, but they should get easier and not

CHAPTER 11. AVOIDING BURNOUT

harder as you gain understanding.

If something seems hard

- you may be making it harder than it is,
- you may not understand the concept fully,
- you may just need more practice,
- or you may not be ready for that concept yet.

If math problems seem hard because you are not getting the correct answers.

Try reviewing the corresponding concept to look for any information you may have missed. Look for similar examples like the problem you are solving to see if you are missing steps. Check your work with the Common Math Errors list in the appendix. Seek clarification by discussing the problems with someone who can help.

If math problems seem hard because it takes so much time to solve them.

New concepts can take a lot of effort to apply at first, and sometimes it takes a lot of practice applying the process of a new concept to several problems before you can apply them efficiently. Just remember that getting good at most things requires a lot of practice and effort. For example, if you have ever taken music lessons then you know that you have to practice playing a song several times before you can play it effortlessly and without any mistakes. The same thing goes with applying a new math concept. It does not have to be hard, it just takes time.

If you think you are just not ready for those problems yet.

Sometimes a new math concept seems hard, not because we can't understand it, but because we are just not ready for that concept yet. Consider the things you need to know in order to understand the current concept. If you are not sure, ask your instructor. Then review those previous concepts before tackling the current one again. If you still do not fully understand, move on anyway. Your mind may just need more time to process the new idea, and you may find that it becomes clear as you learn more advance future concepts. Most courses do not require you to master every concept right away. You should be able to succeed as long as you understand most of what you are learning in the course.

CHAPTER 11. AVOIDING BURNOUT

Questions

1. What is the best way to avoid burn out while taking a math class?

2. What can happen if you procrastinate in a math class?

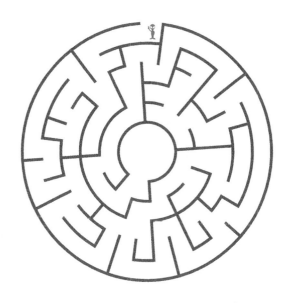

CHAPTER 12 - WORST CASE SCENARIOS

CHAPTER 12. WORST CASE SCENARIOS

> ✖ Pathways
> ───────────────────────────────
> - Participation Scenarios
> - Assignment Scenarios
> - Exam Scenarios

This final chapter is provided for you to reflect on what you have learned and determine solutions to a few worst case scenarios you may encounter in your math class. These scenarios are provided to help you prepare for success even when confronted with adversity and unexpected events.

📍 Participation Scenarios

Participation scenarios are examples of challenges you may encounter while interacting with your course.

Some participation scenarios include:

- **Scenario 1:** The class you want to take is full.

- **Scenario 2:** You signed up for the wrong class, or you no longer wish to take the class.

- **Scenario 3:** You are taking the class online, and you can't find the first assignment.

- **Scenario 4:** You are taking the class online, and you can't remember your username or password to login to the course.

- **Scenario 5:** Most of the assignments are on the computer, and you do not have computer skills.

- **Scenario 6:** You are failing the class.

Each scenario is given with a blank space to write down a possible solution.

Possible solutions for these scenarios are provided in the appendix.

Scenario 1: The class you want to take is full.

Solution:

CHAPTER 12. WORST CASE SCENARIOS

Scenario 2: You signed up for the wrong class, or you no longer wish to take the class.

Solution:

Scenario 3: You are taking the class online, and you can't find the first assignment.

Solution:

Scenario 4: You are taking the class online, and you can't remember your username or password to login to the course.

Solution:

Scenario 5: Most of the assignments are on the computer, and you do not have computer skills.

Solution:

CHAPTER 12. WORST CASE SCENARIOS

> **Scenario 6:** You are failing the class.
>
> **Solution:**

♥ Assignment Scenarios

Assignment scenarios are examples of challenges you may encounter while completing course assignments.

Some assignment scenarios include:

- **Scenario 1:** You do not know how to submit the assignments in the course.

- **Scenario 2:** You are working on the first assignment the night it is due, and you are not sure how to solve some of the problems.

- **Scenario 3:** You forgot to submit an assignment before it was due.

- **Scenario 4:** You are not able to complete the assignments even though you are spending a lot of time studying.

- **Scenario 5:** You receive a poor grade on your first assignment.

Each scenario is given with a blank space to write down a possible solution.

Possible solutions for these scenarios are provided in the appendix.

Assignment Scenario 1: You do not know how to submit the assignments in the course

Solution:

CHAPTER 12. WORST CASE SCENARIOS

Assignment Scenario 2: You are working on the first assignment the night it is due, and you are not sure how to solve some of the problems.

Solution:

Assignment Scenario 3: You forgot to submit an assignment before it was due.

Solution:

Assignment Scenario 4: You are not able to complete the assignments even though you are spending a lot of time studying.

Solution:

Assignment Scenario 5: You receive a poor grade on your first assignment.

Solution:

CHAPTER 12. WORST CASE SCENARIOS

📍 Exam Scenarios

Exam scenarios include examples of challenges you may encounter while completing course exams.

Some exam scenarios include:

- **Scenario 1:** You get stuck on a problem in the middle of taking your exam.

- **Scenario 2:** You do not have enough time to complete the entire exam.

- **Scenario 3:** Your mind goes blank in the middle of taking your test.

- **Scenario 4:** Your computer crashes in the middle of taking an online exam.

- **Scenario 5:** You are taking an online exam the same day it is due, and you realize that you need a webcam, but you do not have one.

- **Scenario 6:** You fail the first exam.

Each scenario is given with a blank space to write down a possible solution.

Possible solutions for these scenarios are provided in the appendix.

Exam Scenario 1: You get stuck on a problem in the middle of taking your exam.

Solution:

Exam Scenario 2: You do not have enough time to complete the entire exam.

Solution:

CHAPTER 12. WORST CASE SCENARIOS

Exam Scenario 3: Your mind goes blank in the middle of taking your test.

Solution:

Exam Scenario 4: Your computer crashes in the middle of taking an online exam.

Solution:

Exam Scenario 5: You are taking an online exam the same day it is due, and you realize that you need a webcam, but you do not have one.

Solution:

Exam Scenario 6: You fail the first exam in the class.

Solution:

CHAPTER 12. WORST CASE SCENARIOS

If there are scenarios you can think of that were not included in this chapter. Consider making a list of your own scenarios and coming up with possible solutions, so you can plan for the worst but hope for the best.

APPENDIX

Worst Case Scenario Solutions

Participation Scenario Solutions:

1. See if you can get on a waiting list for the class. Contact the instructor to see if you can be added to the class. Show up on the first day of class and ask the instructor if you can be added to the class. If possible, attend the class every day of the first week.

2. Withdraw from class as early as possible. Notify the instructor of your plans to withdraw.

3. Look for assignment instructions in an introduction, a syllabus, or an about assignments document or page. Ask the instructor. Ask a fellow classmate.

4. Contact the instructor through a school email or contact the school's student services desk.

5. Ask the student services staff of the school for assistance with computer literacy. Ask the instructor for recommendations.

6. Consider things that may be hindering your success and review appropriate chapters of this book for advice. Ask the instructor for feedback. Consider whether you might need to take a prerequisite math class for the course.

Assignment Scenario Solutions:

1. Check the syllabus for assignment instructions. Look for an about assignments document or page (if online). Ask a fellow student. Ask the instructor.

2. Send questions to the instructor. See if there is a lab available where you can ask questions. Try n free online tutoring service. Review the concept. Look for similar examples to the problems you are trying to solve. Review the "Problem Solving" chapter of this book. Ask a classmate for help. Try to finish homework early in the future.

3. Contact the instructor to see if you can make up the assignment. Turn future assignments in early to avoid this scenario in the future.

4. Review the "Studying Smart", "Using a Textbook", "Learning Big Ideas" and "Solving Problems" chapters of this book. If the recommended strategies do not help, review the "Asking for Help" chapter and ask specific questions about what you do not understand.

5. Analyze the assignment to see what things you need to work on and do differently next time. Make a list of what you can do with the next assignment to ensure a better grade. If necessary, contact the instructor for additional feedback.

Exam Scenario Solutions:

1. Skip the problem and return to it later. Focus first on the problems you can finish quickly.

2. Focus on completing as many questions as you can before you run out of time.

3. Look away from the test for a minute. Take some deep breaths to help your mind relax.

4. Try rebooting and continuing the test. When you are finished, contact the instructor to let them know your computer crashed. See if you can retake the test, if you lot a substantial amount of time.

5. Consider where you may be able to get a webcam locally. Can you borrow a webcam from someone you know? Contact the instructor to see if you can borrow a webcam from the school or get more time to take your test.

6. Analyze your test to see what went wrong. Make a list of things you can do differently with the next test. Review the "Testing Strategies" chapter in this book.

16 Things Active Learners Do

1. Research class before the first day.
2. Purchase required materials before the first day.
3. Overview class materials before the first day.
4. Contact the instructor about concerns..
5. Read the course syllabus on the first day.
6. Take note of important dates.
7. Make a study plan.
8. Review the current class lesson before class.
9. Attend every class.
10. Take notes in class.
11. Ask questions when they do not understand.
12. Ask questions when they want to know more.
13. Participate in class discussions.
14. Complete all graded assignments.
15. Review for exams.
16. Clarify mistakes and learn from them.

15 Things Active Readers Do

1. Do the assigned reading ahead of time.
2. Look over the assigned reading before reading.
3. Pay close attention to headings and emphasized text.
4. Pay close attention to definitions and examples.
5. Consider how the current reading fits into what they already know.
6. Write notes in the margins.
7. Underline, highlight, and make note of important information.
8. Determine key ideas in the section.
9. Write down questions about what they read.
10. Quiz themselves on what they read.
11. Seek clarification as necessary.
12. Read more than once, if necessary.
13. Read aloud.
14. Look for the main ideas in each paragraph.
15. Consider the overall purpose of the reading.

Test Preparation Tips

- Investigate the test
  - Is it a multiple choice test?
  - Is it a short answer test?
  - How many questions will there be?
  - What topics will be covered.
  - What will not be covered?

- Review the main ideas and supporting details covered by the test.

- Practice solving a lot of problems.

- Clarify the things you do not understand.

- Make a cheat sheet of the main ideas.

- Make a practice test.

- Practice stress-relief techniques

- Rest before going to take your test.

Test Taking Tips

- Be prepared - know what you need with you when you take the test and what is allowed.

- At the beginning of the test, write down all the main ideas you can remember (if you cannot use notes).

- Look over the entire test before you begin solving any of the problems.

- Solve the easy problems first.

- If you get stuck on a question, skip it, and return to it later.

- Show work whenever possible.

- If you do not know the answer, write down what you do know.

- Try not to panic. Take deep breaths.

- Check your work for common math errors.

Common Math Errors

- **Copying Error** This happens when we copy a problem and our copy does not match the original problem.

- **Transposing Error** This happens when we write the digits of a number in reverse order.

- **Sign Error** This happens when we drop or add a negative sign in our computation.

- **Legibility Error** This happens when we cannot read our own handwriting.

- **Directions Error** This happens when we misread directions.

- **Computational Error** This happens when we add instead of subtracting, multiply instead of dividing, etc.

- **Number Error** This happens when we write a different number from what we are suppose to write.

- **Calculator Error** This happens when we put the wrong information in the calculator.

Common Test Taking Errors

- **Following Directions**
  - Misreading directions
  - Misunderstanding directions

- **Computation** Knowing how to solve a problem, but making simple mistakes that lead to an incorrect solution.

- **Concept** Not understanding how to solve a problem.

- **Application** Knowing the concept, but not knowing how to apply it to a specific situation.

- **Testing** Making mistakes in how you approach a test.
  - Spending too much time on each question.
  - Going too fast.
  - etc.

- **Studying** Studying the wrong material or spending too much or too little time on test preparation.

Help! I am stuck! Tips

- **Check your work** - Look for common math errors. Did you copy the problem down correctly? Did you make any computational or sign errors?

- **Check the definition** - Did you miss any part of the rules for solving the problem?

- **Check yourself out for a break** - Do something else for a while and return to the problem with a fresh perspective.

- **Check with a different pair of eyes** - Have someone else check your work.

- **Ask for help** - If all else fails, find someone who can help.

Made in the USA
Las Vegas, NV
26 September 2021